EARTHQUAKES TO VOLCANOES

Design	David West Children's Book Design
Designer	Keith Newell
Editorial Planning	Clark Robinson Limited
Picture Researcher	Emma Krikler
Science Editor	Catherine Warren
Illustrator	Ian Moores
Consultant	Michael Stuart, geography teacher

© Aladdin Books 1992

First published in
the United States in 1992 by
Gloucester Press
95 Madison Avenue
New York NY 10016

Clark, John Owen Edward.
    Earthquakes to volcanoes : project with geography / by John Clark.
      p. cm. -- (Hands on science)
    Includes index.
    Summary: Discusses plate tectonics and the causes of earthquakes and volcanoes, the features and formation of other geographical areas, and possible future changes in the earth's surface. Features projects throughout.
    ISBN 0-531-17316-X
    1. Plate tectonics--Juvenile literature. 2. Earthquakes--Juvenile literature. 3. Volcanoes--Juvenile literature. 4. Plate tectonics--Experiments--Juvenile literature. [1. Plate tectonics. 2. Earthquakes. 3. Volcanoes.] I. Title. II. Series.
QE511.4.C57   1992
551.1'36--dc20   91-35076   CIP   AC

All rights reserved

Printed in Belgium

HANDS·ON·SCIENCE

EARTHQUAKES TO VOLCANOES

John Clark

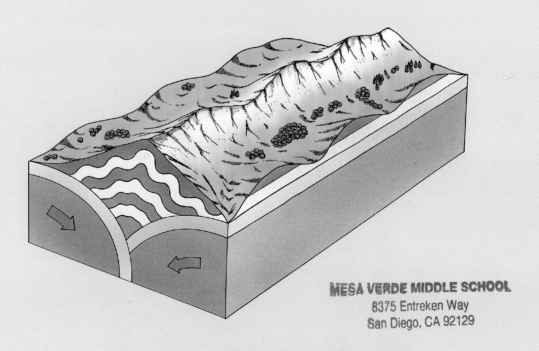

GLOUCESTER PRESS
New York · London · Toronto · Sydney

CONTENTS

EARTH'S INTERNAL STRUCTURE	6
EARTH'S CRUST	8
EARTHQUAKES	10
VOLCANOES	12
HOW ROCKS ARE FORMED	14
FOLDS AND FAULTS	16
FORMATION OF MOUNTAINS	18
OCEANS	20
WIND EROSION	22
WATER AND ICE EROSION	24
SEA EROSION	26
UNUSUAL LANDSCAPES	28
BELOW THE EARTH'S SURFACE	30
GLOSSARY	31
INDEX	32

Earthquakes and volcanic eruptions are two forces that help to shape the earth. This book takes a closeup look at our planet's structure and how its surface changes. It tells you about rocks, how they are formed, and how earthquakes, volcanoes and erosion turn them into valleys and mountains. There are "hands on" projects for you to try, which use everyday materials and equipment.

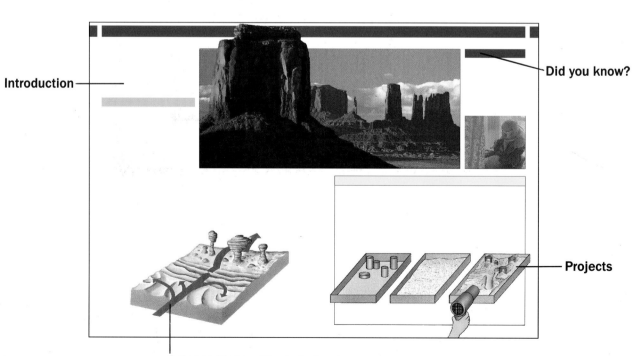

Introduction

Did you know?

Projects

Science ideas with photographs and diagrams

INTRODUCTION

Over a period of a few hundred years, the surface of the earth appears not to change. But during the millions of years of the earth's existence, it has changed many times. Earthquakes and volcanoes alter the face of our planet. But there are also slower, long-term processes at work.

The solid surface of the earth — the crust — consists of gigantic slabs of rock called plates that float on the molten magma beneath. As they move, their edges slowly collide into each other, pushing up mountains and releasing molten rock to form volcanoes. Folding and cracking of the rocks at the surface also change the landscape.

As well as these forces from within the earth, outside factors come into play. Wind, rain and frost erode rocks and break them into fragments, which rivers carry to the sea. The sea itself erodes coastlines as waves pound the shore.

Erosion has worn away soft rocks to leave jagged pinnacles of hard rock.

EARTH'S INTERNAL STRUCTURE

The earth is a vast globe, 8,000 miles across, spinning on its axis as it orbits around the sun. Seen from space, it looks mostly blue, with patches of white where clouds hide the oceans and land below. But underneath that fairly calm exterior is a boiling hot caldron of molten rock and metal which makes up the inner earth.

FORMATION

The earth was formed about 4.5 billion years ago at the same time as the other planets and the sun. As the earth took shape from spinning masses of gas and dust, the heavier materials sank to the center. Lighter solids, liquid and gas formed a layer around the outside.

Heat from the center of the earth kept the surface a bubbling mass of molten rock for millions of years. Then, about four billion years ago, it began to solidify into a rocky crust dotted with thousands of active volcanoes. Rain poured down while violent thunderstorms raged over the whole planet.

About 1.5 billion years later, blue-green algae began to produce oxygen changing the composition of the atmosphere. The crust settled into enormous plates floating on the fiery material beneath. These are the same moving plates that form the crust today.

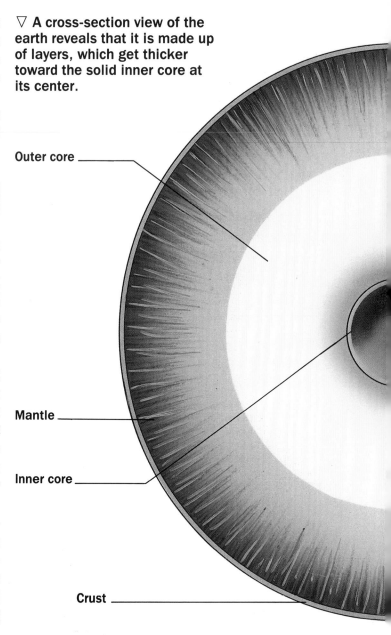

▽ A cross-section view of the earth reveals that it is made up of layers, which get thicker toward the solid inner core at its center.

Outer core

Mantle

Inner core

Crust

← PRECAMBRIAN

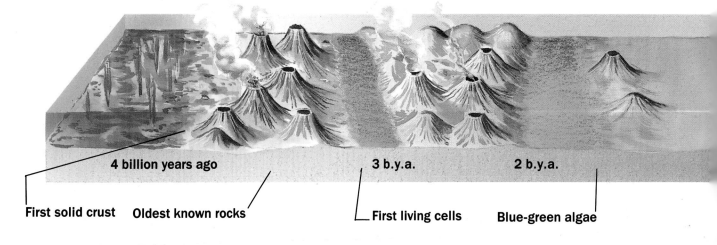

Continents and oceans form

Buildup of oxygen in atmosphere

4 billion years ago

3 b.y.a.

2 b.y.a.

First solid crust   Oldest known rocks

First living cells

Blue-green algae

FROM CORE TO SURFACE

The land masses of the earth consist of solid rock. The ocean beds also are made of rock. But the earth is not solid all the way through. The land and the ocean beds form the outer crust, and below this is the solid mantle. Beneath that is a deep liquid layer of molten iron, with a central core of solid iron and nickel.

The earth has a layered structure like an onion, but only the center and the outer crust are hard. About one-third of the crust is dry land. The rest is covered by the oceans. Both on land and on the ocean floors the crust moves. This causes earthquakes and gives rise to volcanoes where liquid magma (molten rock) forces its way to the surface.

DID YOU KNOW?

The pie chart (below) shows that the most abundant element in the earth's crust is oxygen (46.6 percent by weight). Most of it is combined with the second most abundant element, silicon, in the form of silicate rocks. Aluminum is the most common metal.

△ Much of the earth is covered by water, with the continents resembling giant islands. Above the surface is the atmosphere, which extends upward for over 300 miles before it fades into outer space.

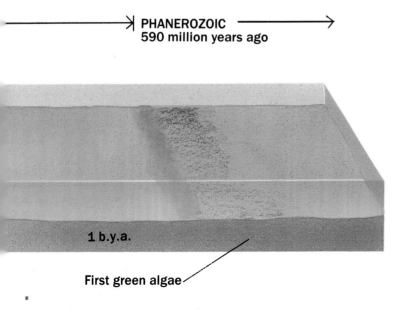

PHANEROZOIC
590 million years ago

1 b.y.a.

First green algae

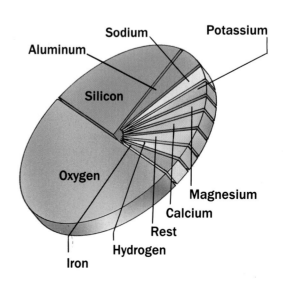

8 EARTH'S CRUST

The earth's crust is a rocky skin formed of huge interlocking plates, which are about 25 miles thick in areas occupied by the continents. But beneath the oceans the crust is only about 5 miles thick. The crust is constantly changing, as currents of molten material deep in the earth keep these plates on the move.

DRIFTING CONTINENTS

Each of the continents is anchored to one or more of the earth's crustal plates (see map opposite). The continents have not always been in their present position. About 200 million years ago they were grouped together into a single supercontinent called Pangaea. But as molten material from inside the earth forced its way up between the plates, new crust was formed and the plates began to move apart, breaking up Pangaea into the continents we know today.

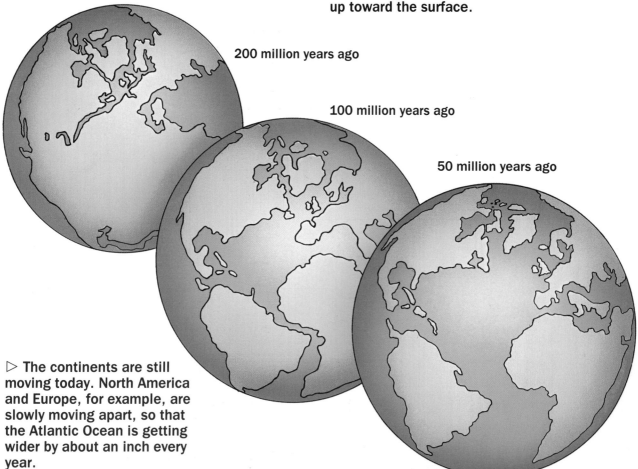

Irregular convection currents in mantle

Moving plates meet; one sinks beneath the other.

Sinking plate melts back into mantle.

△ Heat generated deep in the earth causes convection currents in the mantle as the hotter material slowly moves up toward the surface.

200 million years ago

100 million years ago

50 million years ago

▷ The continents are still moving today. North America and Europe, for example, are slowly moving apart, so that the Atlantic Ocean is getting wider by about an inch every year.

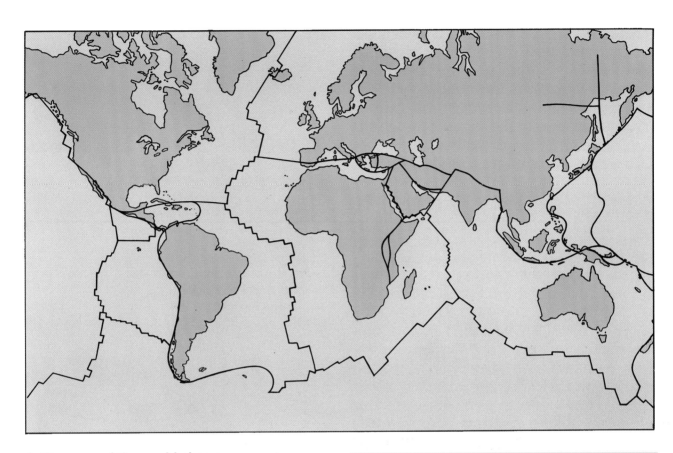

△ The map of the world shows the boundaries between the main crustal plates. Volcanoes, earthquakes and mountain ranges are characteristic features along these boundaries.

Present day

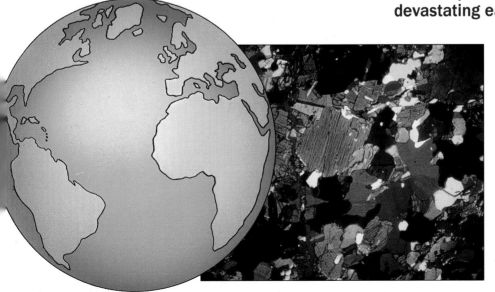

PLATES IN MOTION

Sometimes at the edges of moving plates violent and spectacular phenomena occur. Plates push against one another with enough force to throw up great mountain chains. The tremendous pressure in the mantle squeezes molten material upward between the plates to form volcanoes. Plates under great tension may rupture and snap into a new position causing devastating earthquakes.

DID YOU KNOW?

The rocks that make up the earth's crust are made of crystals. Some, such as quartz and sand, are obviously crystalline. Others only reveal their crystal structure when polished, sliced thinly and viewed with the aid of a powerful microscope (left).

10 EARTHQUAKES

Earthquakes are among the most destructive and terrifying of natural events. They vary in strength from minor tremors to violent vibrations that topple buildings and open up huge cracks in the ground. Fortunately, most take place beneath the sea, although even then they may cause tidal waves that can reach the coastline.

WHY THEY HAPPEN

Earthquakes will arise at plate boundaries where two plates are pushing against each other under the pressure of continental drift (see page 8).

Rocks have elastic qualities and they can absorb this pressure for hundreds or even thousands of years. But eventually the strain is too great and the rocks rupture and jerk into a new position releasing all the pent-up energy in the form of an earthquake.

Vibrations spread out from the center or "focus" of the earthquake causing the ground to shudder violently. The point on the surface immediately above the focus is called the epicenter. Here the earthquake is most severe and damaging.

△ Loss of lives, wrecked buildings and cracked roads followed a devastating earthquake in Mexico City in 1985.

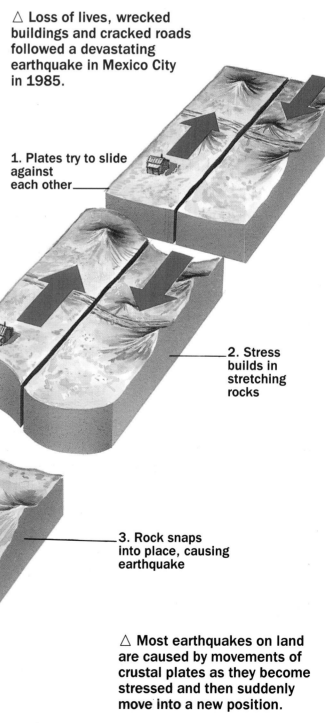

1. Plates try to slide against each other

2. Stress builds in stretching rocks

3. Rock snaps into place, causing earthquake

△ Most earthquakes on land are caused by movements of crustal plates as they become stressed and then suddenly move into a new position.

MEASURING EARTHQUAKES

The strength of an earthquake can be measured by instruments called seismographs. Most have pens that trace wavy lines on a chart as a recording of the event, and these tracing are called seismograms. These instruments have a very heavy weight which, although suspended like a pendulum, tends to stay still while the rest of the device is moved rapidly by earth tremors.

The effects of an earthquake in a particular place is described by the Mercalli scale, which ranges from I to XII. The Richter scale, a logarithmic scale which has a maximum of nine, measures an earthquake's strength at its focus.

△ Earthquake tremors are caused by shock waves traveling rapidly through the ground. A seismograph mirrors these tremors as a series of wavy lines on a chart.

MAKE A MODEL OF AN EARTHQUAKE

To model the sideways movement of rocks that causes most earthquakes, you will need two rectangular blocks of wood and a sheet of paper. Put the blocks side by side and glue the paper to them, being careful not to let glue get in between the blocks. Now stand some thumbtacks upside down on the paper. Finally imitate the forces of plate movement by pushing on the ends of the blocks in opposite directions. At first the paper will resist your pushing, but then it will suddenly tear, splitting apart the blocks and toppling the tacks.

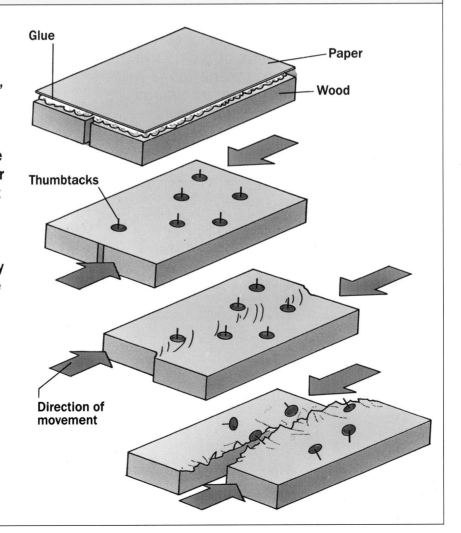

VOLCANOES

A volcano forms when a hole or crack in the earth's crust lets magma force its way through. On land, ash and smoke may first explode into the air. Then magma, called lava above the surface, oozes out of the volcano. Undersea volcanoes cause great clouds of steam to bubble to the surface.

VOLCANO CREATION

Underneath a volcano there is a cavity of molten rock called a magma chamber within the mantle. It forms below a weak point in the crust, possibly below a midocean ridge where crustal plates are moving apart. The magma is under pressure and less dense than the mantle so it gradually rises, often up cracks, or fissures, in the crust. Gases are produced and eventually the pressure builds up so much that they blast a way to the surface.

At this stage, the volcano belches out gases, dust and fragments of rock. Lava flowing from a crack in a plane forms a lava plateau. Lava that piles up around the opening (called a vent) forms a typical cone-shaped mountain.

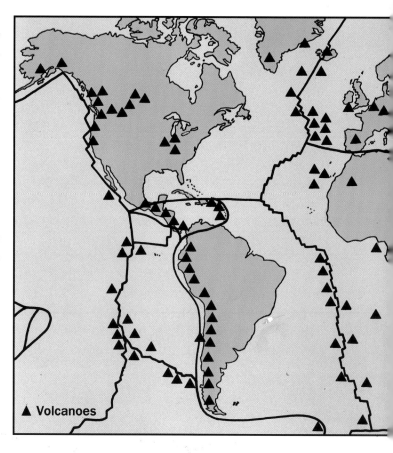

▲ Volcanoes

△ There are more than 2,000 volcanoes under the sea. Most land volcanoes are in mountain chains, such as the South American Andes.

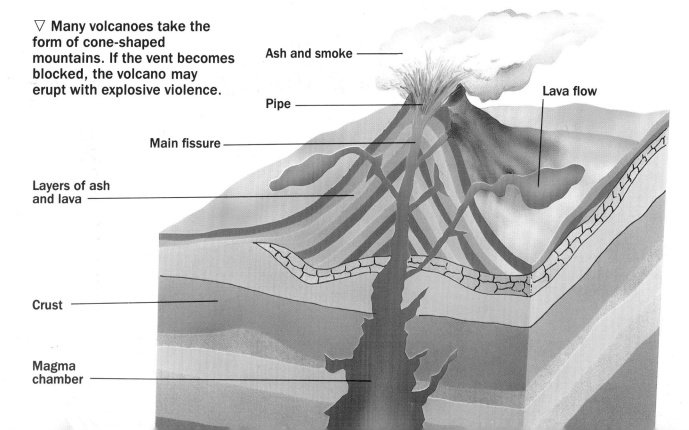

▽ Many volcanoes take the form of cone-shaped mountains. If the vent becomes blocked, the volcano may erupt with explosive violence.

DESTRUCTION

An unexpected, sudden volcanic eruption can shower the surrounding area with burning hot ash and cinders. Such an eruption of Vesuvius in Italy wiped out the roman city of Pompeii in AD 79. A much more recent disaster occurred on May 18, 1980 when a whole side of Mount St. Helens, a volcano in the state of Washington, was blasted away. A white-hot cloud of gas and powdered magma smothered everything within 5 miles of the mountain. The explosion was estimated to equal the power of 500 atomic bombs. Another danger from volcanoes is fire. This is caused by molten rock hurled into the air or lava flows down hillsides setting fire to things.

▽ Thousands of sparks and lumps of molten rock hurled high in the air create a huge fireworks display as a volcano bursts into life.

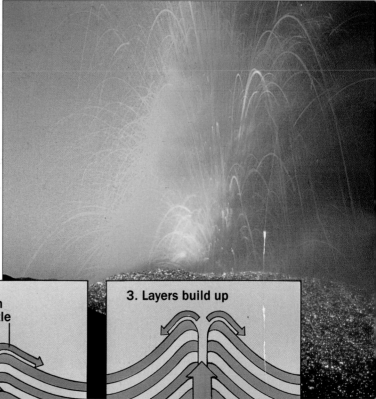

WHERE THEY ARE FOUND

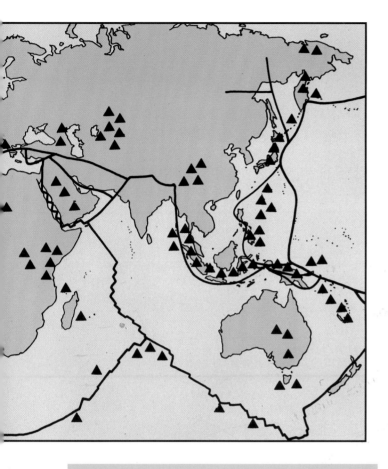

Over half the world's volcanoes arise in a belt around the Pacific Ocean called the Ring of Fire. Here plate edges overlap and are dragged back into the mantle. The old crust melts and immense pressure can force magma back to the surface. Along midocean ridges the crust is thin and weak and magma rises to form a line of volcanoes.

▽ Layers of ash welded with solidified lava shape the mouth and sides of the vent.

1. Lava flow
2. Layers of ash and lava settle
3. Layers build up

HOW ROCKS ARE FORMED

The earth's crust is made of rocks, and the rocks are made up of minerals, which are chemicals formed in the earth. There are three types of rock: igneous rock, metamorphic rock, and sedimentary rock. Each type contains characteristic minerals. Rocks are being created and destroyed all the time in the "rock cycle."

IGNEOUS ROCKS

Igneous rocks are formed when molten magma from deep within the earth cools and solidifies to form the crust. When magma is forced through the vent of a volcano as lava, it also solidifies to rock as it rapidly cools. Igneous rock is classified by its silica content. Those rich in silica, such as granite, are light in color and are called "acidic" rocks. "Basic" rocks, low in silica, are dark in color.

△ These mountains are made of granite, a hard igneous rock formed when molten magma from the earth's mantle cooled to form part of the crust.

Granite tor (Igneous)

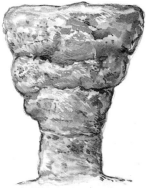

1. Igneous rock forms as lava is deposited and cools around the volcano.

2. Rocks are also brought to the surface by the movement of the earth's crust.

3. Metamorphic rock is formed from igneous and sedimentary rocks under heat and pressure deep in the earth's crust.

The rock cycle
▷ Different types of rock are formed from existing ones in a continual process known as the rock cycle.

METAMORPHIC ROCKS

Metamorphic rocks are formed when conditions within or on the surface of the earth change other rocks. Deep underground, heat and pressure can alter minerals in rocks, for example changing the sedimentary rock limestone into its crystalline metamorphic counterpart, marble. Pressure under mountains produces layered rocks such as slate which cleave into thin plates if damaged.

SEDIMENTARY ROCKS

At the earth's surface, wind and rain cause erosion. Rocks are broken down into small grains which are washed away by rivers to form sediments. As layers of sediment build up, those underneath are squeezed and the pressure turns them into sedimentary rocks. Limestone and sandstone are typical sedimentary rocks. Earth movements or erosion may expose them at the surface.

Marble

Claystone, a sedimentary rock, metamorphoses into slate.

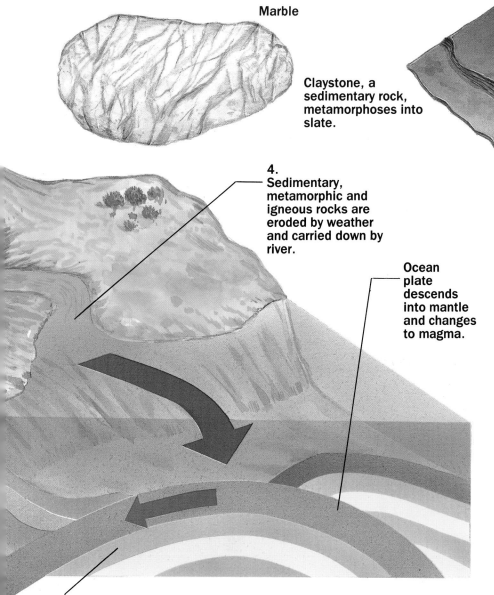

4. Sedimentary, metamorphic and igneous rocks are eroded by weather and carried down by river.

Ocean plate descends into mantle and changes to magma.

5. Sedimentary rocks eventually form due to pressure from continuous layers of sediment.

DID YOU KNOW?

The remains of dead animals may be buried in the sediments that later turn to sedimentary rock. When they do, the hard parts (such as bones and shells) may be replaced by rock-forming minerals and become fossils. Leaves, pollen grains, and even dinosaur eggs also form fossils.

FOLDS AND FAULTS

Sedimentary rocks are laid down originally in horizontal layers, or strata. But the pressure of crustal plate movements may push the layers and bend them into folds. Alternatively, the rocks crack under the pressure, and breaks, called faults, appear. The results are hills, mountains, valleys, and gorges.

COLLIDING PLATES

Most folds are caused by movements of the huge plates that form the earth's crust (see page 9). If the resulting sideways movement is small, a range of gently rolling hills may be formed. Large movements push the strata into loops (called anticlines) and troughs (synclines). These may form high mountains and deep valleys. Then over thousands of years, wind and rain erode the mountains, exposing the tilted strata which appear as bands of rock.

Folding also occurs when a plug of igneous rock or crystalline salt is forced upward from below. The overlying layers of sedimentary rock are pushed up and arched into a dome. Often the strain on the rocks at the roof of the dome cracks them into faults. The plug may even break through at the top of the dome. Salt-plugs are important because they may trap natural gas and oil when they are below impermeable rocks.

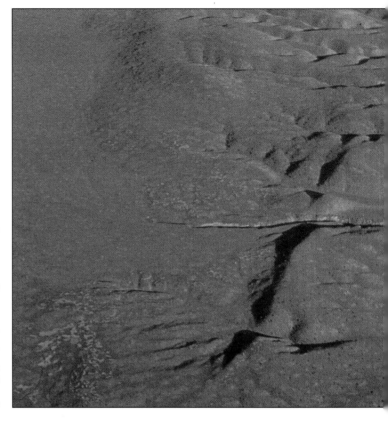

△ The San Andreas Fault in California (ridge running top to bottom in picture) is subject to sideways movements causing earthquakes, like the devastating San Francisco quake in 1989.

▽ When one of the earth's huge plates slides below a continental one (1), the edge of the upper plate folds (2). This is one of the chief ways that mountains are formed.

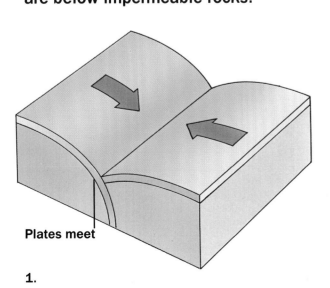

1.

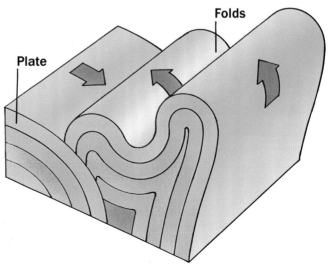

2.

CRACKING ROCKS

When regions of rock are pulled apart or pushed together, a block of rock between them may fall between what are called normal faults. If the angle of the fault is very shallow, one block may rise over the next in a reverse fault.

Sideways movements of large blocks of rock cause a transcurrent fault, which can cause earthquakes (see page 10).

▽ A falling block (1) forms a valley. Rising pressure (2) can form hills. Sideways shifts (3) can cause earthquakes.

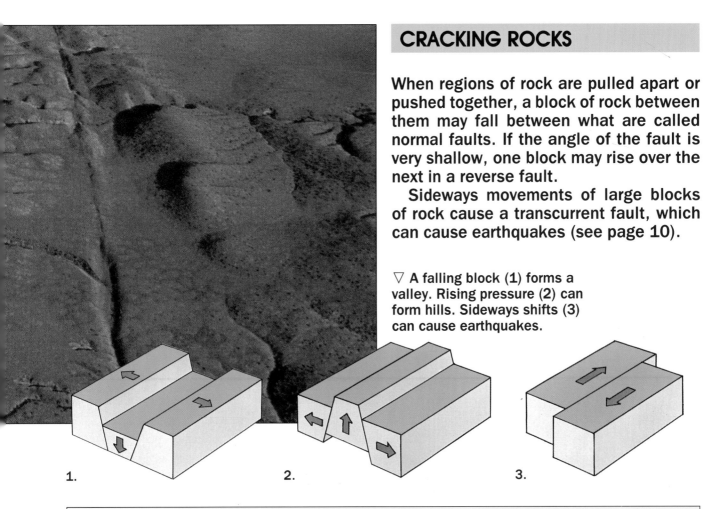

1.   2.   3.

HOW FAULTS WORK

Take three pieces of thick cardboard. Cut two of the pieces at an angle of 45 degrees. Cut the other piece into a flat-topped triangle. You can see the shapes this will make in the drawing. Put the pieces of cardboard onto a flat table. In turn, try the three experiments shown in the diagram, each time pushing the two outer pieces of cardboard toward each other. The outer pieces represent moving crustal plates. You can see what happens when there are different faults arrangements.

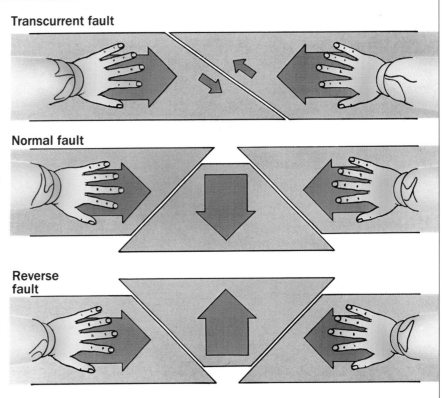

Transcurrent fault

Normal fault

Reverse fault

FORMATION OF MOUNTAINS

Mountains are chiefly formed by folds and faults (see pages 16-17). The action is particularly violent when two crustal plates collide and the edge of one is crumpled into chains of fold mountains. Faulting can give rise to block mountains, and pressure from rising magma can push up dome mountains.

FOLD MOUNTAINS

At the edge of a continent where two plates meet, the continental plate rises over the denser oceanic plate. The relentless pressure on the plates causes the continental plate to buckle, throwing up a mountain range. It also scrapes off the sediment from the oceanic crust, folding it into more mountains. This process formed the Andes in South America.

When two continental plates collide, one does not rise above the other. The edges continue to push into one another and the tremendous force buckles the two plates and folds them into huge mountain ranges. Examples of this include the Himalayas in Asia and the Alps in Europe. These two mountain ranges are relatively young and continue to rise even today.

△ The Himalayas, a range of fold mountains that extend along the northern part of the Indian subcontinent, include Mount Everest (29,029 feet) the world's highest peak.

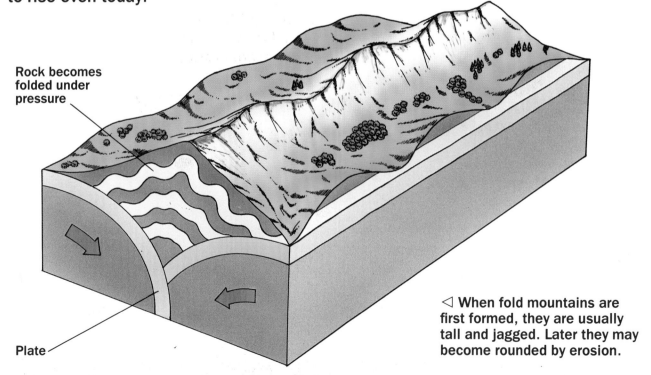

Rock becomes folded under pressure

Plate

◁ When fold mountains are first formed, they are usually tall and jagged. Later they may become rounded by erosion.

BLOCK MOUNTAINS

The tallest mountain in Africa is Mount Kilimanjaro. Although it lies close to the Equator, it is so tall (the larger of its twin peaks towers 19,341 feet above the surrounding plane) that its top is permanently covered in snow more than 200 feet deep. It is an extinct volcano, one of a chain of block mountains that form the eastern rim of the Great Rift Valley. They were formed when the block that forms the valley floor sank because of extensive faults.

△ The Great Rift Valley extends for much of the length of eastern Africa, formed when blocks of rock sank between faults.

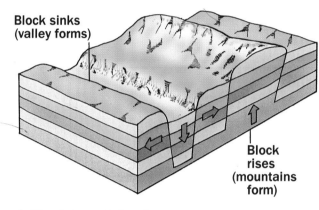

△ Block mountains form alongside a rift valley, and may be pushed higher by the pressure of magma beneath.

▽ Dome mountains, formed by pressure from below, begin as rounded hills, but can be shaped by wind and weather.

DOME MOUNTAINS

Sometimes a great mass of magma beneath the earth's crust gathers to form a magma chamber. As the chamber grows it exerts great pressure on the layers of rock above it. These layers are pushed up into great arcs of rock. The result is one or more dome mountains, such as the Black Hills in South Dakota, United States. The existing Black Hills, up to 3,940 feet tall, are stubs of hard rock that were left when their softer outer layers were eroded away.

▽ Erosion has stripped away the softer rocks of the Black Hills of Dakota, leaving scores of rounded dome mountains.

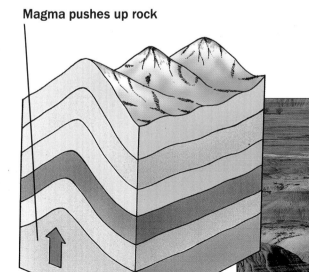

OCEANS

About three-quarters of the earth's surface is covered by seas and oceans. The oceans occupy huge basins averaging 2.5 miles deep, sitting on a layer of the earth's crust. But the crust under the oceans is only 4-5 miles thick, compared to the 25 miles thickness of crust below the land.

UNDERWATER CRUST

Just as the crust forming the land surface is slowly moving and changing, the rocks that form the ocean floor are also on the move. Many of the crustal plate boundaries (see pages 8-9) are along the edges of oceans or form midocean ridges. At these ridges, molten magma wells up from below the crust and forces the sides of the ridge apart. This movement, called seafloor spreading, is gradually widening the world's major oceans. In other places, one plate may dip below another to form a deep ocean trench. The Mariana Trench, in the Pacific Ocean, is 23,000 feet deep (extending 36,000 feet below sea level). Both types of plate boundary may be associated with underwater volcanoes, which may grow tall enough to form chains of new islands or to support coral reefs.

△ A world map of the oceans shows how the ridges and trenches correspond to plate boundaries (see page 9).

▽ The ocean floor is not flat but has as many cliffs, volcanoes, mountains and deep valleys as any dry landscape.

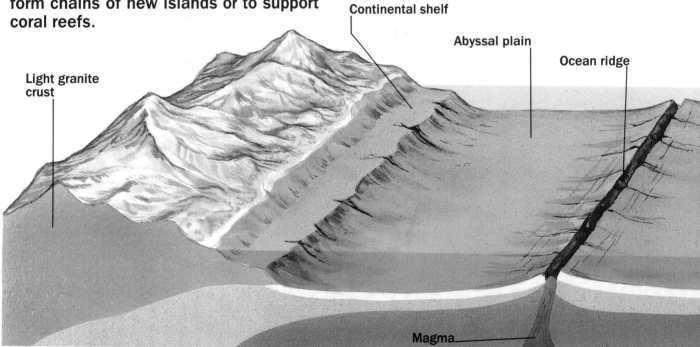

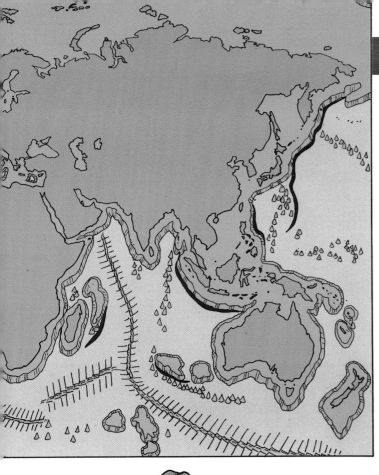

CORAL ISLANDS

Coral is formed from the rocky skeletons of millions of small sea creatures called polyps. They need sunlight and can live only in shallow water. A coral island in the middle of an ocean must therefore grow on or around something — usually an extinct volcano. Movement of the oceanic plate causes the volcano to "sink" as its piece of crust moves into deeper water. At the same time, coral growths first form a fringing reef, then a barrier reef and finally an atoll as the volcanic peak disappears altogether. The atoll takes the form of a circular reef with a central lagoon, made shallow by sediments of coral that have been worn away.

◁ The diagrams show the stages from fringing reef to guyot formation. The atoll (below) is one of many in the Pacific Ocean.

Fringing reef

Barrier reef

Coral atoll

Guyot

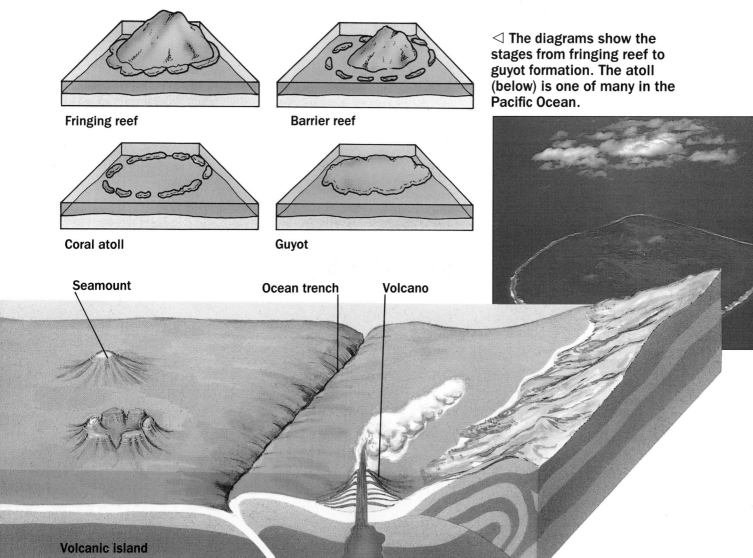

WIND EROSION

Major earth movements give the landscape its basic form. But it is then shaped by various kinds of erosion, mainly the action of wind, water, and ice, and the sea. Wind erosion has its greatest effect in dry regions, particularly deserts, where windblown sand carves rocks into strange shapes.

DESERT LANDSCAPES

In hot, sandy deserts, a steady wind blows the sand into parallel lines of transverse dunes, which resemble sea waves that have been frozen to show their shape. If the wind is turbulent, it blows the sand into seif dunes and into crescent-shaped barchans as shown below.

The wind also makes coarse grains of sand and particles of soil bounce along within a foot of the ground. When these fast-moving particles hit a pillar of rock, they erode it away near the ground to shape it into a top-heavy feature known as a pedestal.

Small boulders on the ground are worn smooth on the side facing the prevailing wind. If they overbalance a different side is sand-blasted, and then another one, until eventually they form structures known as dreikanters. Larger rock formations are carved into rounded outcrops called inselbergs.

△ The features of this desert landscape have been caused by the wind as it carries along sand that carves the rocks into fantastic shapes like these mesas (flat-topped hills) and buttes (narrow columns).

▷ The shapes of sand dunes depend on the direction of the wind and whether it is steady or turbulent. Windblown sand also erodes the rocks into distinctive shapes.

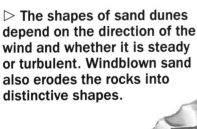
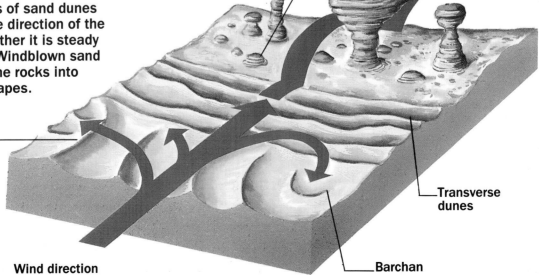

Seif dunes

Wind direction

Sand-blasted rocks

Pedestal

Transverse dunes

Barchan

DID YOU KNOW?

Sandblasting is an industrial process that uses jets of high-pressure air to blow sand for cleaning metal components and buildings. It is similar in action to the natural processes that erode rocks in dry desert lands.

HOW TO FORM A DESERT LANDSCAPE

You can imitate the action of wind erosion and make your own desert landscape. First take a shallow box or tray and stick in place some pillars made from cylinders of modeling clay (1). Next sprinkle sand into the box until the pillars are covered (2). Then, with the tray on sheets of newspaper, use a hair drier to act as the wind (3). Carefully blow away the sand and watch it form dunes and gradually reveal the buried pillars.

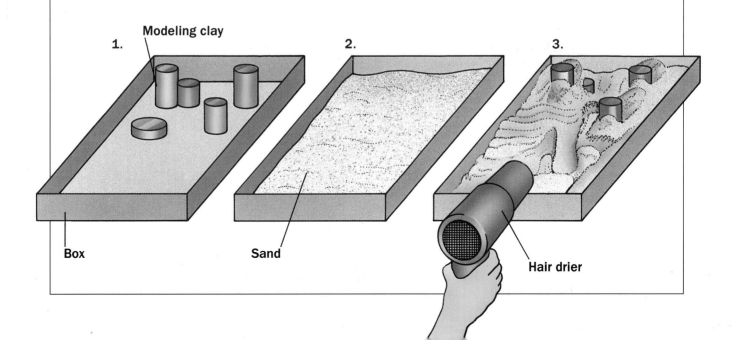

1. Modeling clay — Box
2. Sand
3. Hair drier

WATER AND ICE EROSION

Water unleashes one of the most powerful forces of erosion. Running water washes away soil and rock particles and carries them hundreds of miles along river beds. Frozen water, in the form of ice, breaks up rocks or gouges a path for itself when it slowly flows down a mountainside as a glacier.

VALLEYS

Most valleys result from the action of running water fed by rain and springs. Young mountain rivers splash over cliffs as waterfalls as they carve their course. Rivers broaden and meander as slopes get less steep and become wide, shallow valleys. Oxbow lakes may form when meanders become so extreme that they cut themselves off as the river continually erodes its own banks. Nearing the sea the river may fan out to form a delta between marshes or mudflats made from silt carried down by the river.

△ Many of the features of this landscape have been formed by the river as it flows along the valley.

▷ In its upper course, a fast-flowing young river cuts gorges and canyons. Lower down it slows and becomes shallower. At its mouth, there may be marshes and a delta.

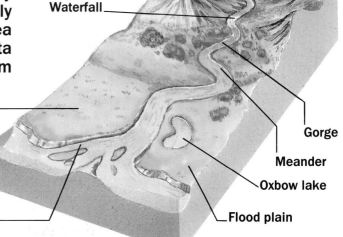

THE PATH OF WATER

You might expect water to flow down a gentle slope in a straight line. But you can demonstrate that this is not so by making a slope of sand in a box and trickling water near the top using a spoon. Notice the path the water takes. Try pouring the water faster from a jug and see the different path it takes.

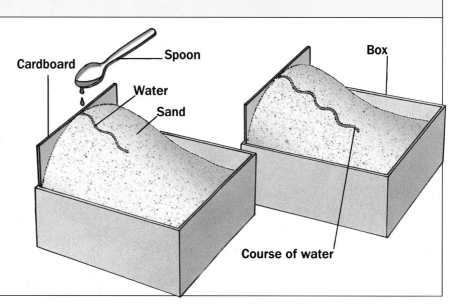

GLACIERS

Glaciers are like frozen rivers. But they are not rain water that has frozen; they are formed of compressed layers of snow that get squeezed under their own weight to form ice. Very slowly, a glacier flows down a mountainside, carving a valley for itself and pushing rock debris in front. If a change in climate makes the ice melt, as it did after the last great Ice Age about 10,000 years ago, rounded valleys reveal where the glaciers used to be.

▽ Most valley glaciers move less than 10 inches a day, pushing along a pile of rocks called a moraine.

△ Sometimes the U-shaped valley left after a glacier has melted becomes flooded by a rise in sea level. It then forms a fiord, as in Scandinavia and New Zealand.

THE POWER OF ICE

Unlike most solid substances, ice expands as it gets colder. This is why water pipes sometimes burst in a severe winter; the force of the expanding ice is enough to rupture metal. The same force can also break even the hardest rock. In winter, low temperatures can freeze rain water that has entered cracks and fissures in rock. If the weather gets even colder, the ice expands and cracks the rock to pieces.

In areas such as the northern tundra, where it is below freezing for most of the year, frost has the effect of moving and mixing soil particles. Expanding ice crystals below stones tend to lift them to the surface, where they gather in grooves and cracks that form between flat domes of frozen soil.

△ Landscapes formed by the action of ice have many characteristic features. The diagram (right) shows how ice can split rocks apart.

SEA EROSION

One of the features of a landscape that changes most quickly is the shoreline. Even the hardest rocks cannot withstand the ceaseless pounding of the waves as wind and current drive them onto the shore. Cliffs crumble or become eroded into stacks, caves and caverns. Sandy beaches may be washed away.

COASTAL FEATURES

Few coastlines are straight when first formed. Most have inlets and bays separated by headlands that jut out into the sea. But then the waves get to work.

As the wind blows waves toward a headland, they swing around and hit it on the sides. This has the effect of gradually wearing away the headland. Sometimes a stack or an arch of rock is left where the sea breaks through a narrow headland. Often caves are gouged out in the cliffs, particularly if they are made of a soft rock such as chalk.

But even cliffs are not permanent. The pounding of the waves, often carrying rock particles of many sizes in the swirling water, cuts a notch on the waterline at the base of the cliff. Eventually the rock above the notch splits away and tumbles into the sea. Buildings on badly eroded coastal cliffs, may also fall as the cliffs are gradually eaten away. Cliffs might erode by as much as seven feet every year.

△ Waves often erode chalk cliffs into strange shapes. A stack or pillar can be cut off a headland, or an arch formed where a cave is parted from the main cliff.

▷ Caves are formed in cliffs where the sea pounds against them. A powerful Atlantic wave exerts a force of about ten tons per square yard and may carry shingle and pebbles that batter the cliff.

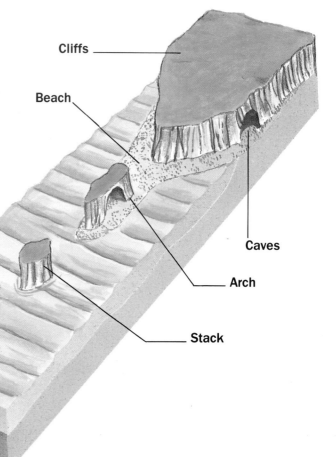

DID YOU KNOW?

The relentless pounding of the sea can reduce rocky cliffs into sandy beaches. The waves continually batter the coastline slowly eroding the cliffs. The rocks chip and crumble into the sea where they are constantly hurled against each other and worn down until they become small, smooth pebbles and eventually sand.

MAKE YOUR OWN COASTLINE

Stick some cylindrical blocks of modeling clay in one half of a deep plastic tray, and cover them with sand as in the diagram. Carefully pour water into the other half of the tray. Now use a piece of plastic or stiff cardboard to make waves, and see how your coastline is gradually eroded away.

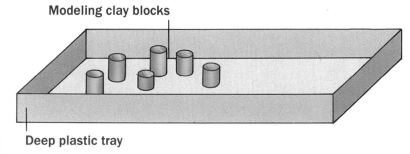

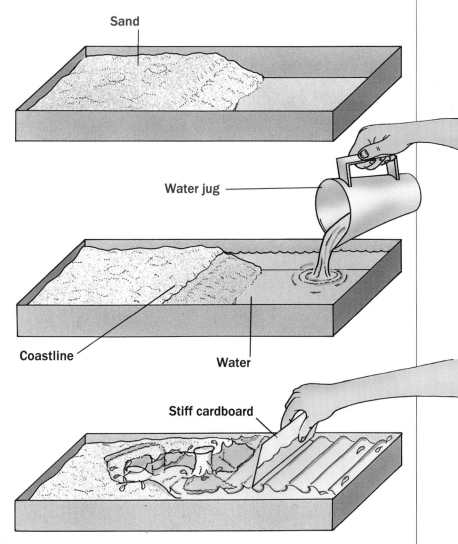

28 UNUSUAL LANDSCAPES

Earth movements and surface erosion are not the only forces that shape the landscape. Some forms of erosion alter regions hidden underground. Occasionally, meteorites from outer space crash onto the earth and form large craters. And human activities such as mining also shape the land.

UNDERGROUND EROSION

When rain water dissolves carbon dioxide from the air it forms carbonic acid which attacks limestone. It carves out surface gullies and gouges out caves where streams flow through the limestone underground. The process takes thousands of years. Water that seeps through limestone contains dissolved calcium carbonate. As the water drips from the roof of a cave, the calcium carbonate comes out of solution and forms rocky stalactites that hang like icicles from the roof and stalagmites that stand like small spires beneath.

△ The rocky spikes hanging from the cave roof are stalactites. The upward-pointing ones are stalagmites. Both were formed from dripping water containing calcium carbonate.

▷ This meteorite crater in Arizona, is about one mile across and more than 558 feet deep.

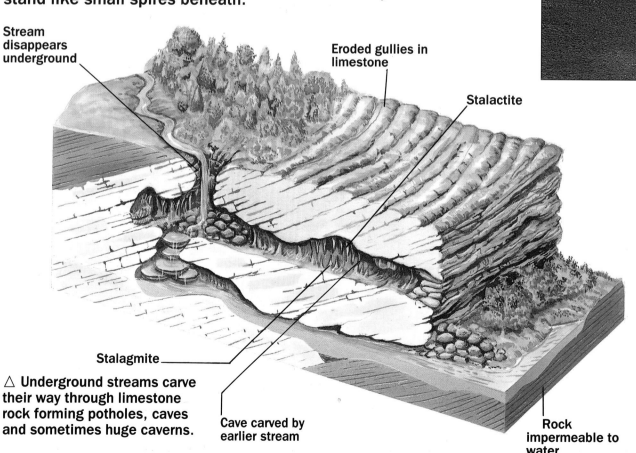

△ Underground streams carve their way through limestone rock forming potholes, caves and sometimes huge caverns.

METEORITE CRATERS

Meteors are collections of rocky or metallic particles often mixed with ice that orbit the sun. If the earth passes through their orbit, they may enter the earth's atmosphere and burn up as shooting stars. Very rarely a large meteor does not burn up completely, but crashes to the ground as a meteorite. The impact can cause a huge explosion and leave a giant crater. In forested areas, it can flatten all the trees for many miles around.

HUMAN INFLUENCE

There are various ways that human activities can change the landscape — usually for the worse. Large strip mines for bauxite, coal and copper can carve away whole mountains. Chopping down rainforests can lead to rapid erosion as the soil is washed away, leaving a rocky wasteland. In dry areas, the removal of shrubs and trees allows winds to blow sand over farmland, thus changing it into a barren desert.

△ Strip mining can permanently alter the shape of the landscape. This enormous hole is a copper mine in Montana, United States.

MAKING A CRATER

This project is messy, so wear old clothes and do it outdoors. First mix some plaster with water to make a very thick paste. Spread it in the lid of a box (1) and drop a stone or a ball of modeling clay into the middle of the wet plaster (2). Carefully remove the "meteorite" (3) and let the plaster set. You will have a good model of a meteorite crater.

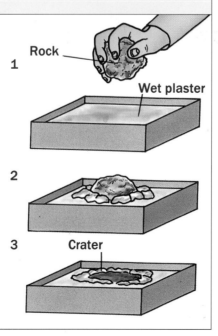

DID YOU KNOW?

Polar regions have rarely had ice caps throughout the history of the earth. Warm ocean currents penetrated into these high latitudes and kept the Arctic Ocean ice-free. But for the past few million years Arctic waters have been almost entirely surrounded by land masses, blocking warm currents. A thin ice sheet consequently formed over the Arctic Ocean creating a unique landscape where there is no solid ground.

BELOW THE EARTH'S SURFACE

In prospecting for oil below the earth's surface, scientists create miniature earthquakes using explosives. By timing the shock waves using microphones set at increasing distances from the explosion, they can work out the rock structure beneath the surface. This helps them determine the likelihood of finding oil.

A known earthquake zone in California lies astride the San Andreas Fault. American scientists have placed satellites in stationary orbit immediately over the area. Laser beams at precisely mapped locations on either side of the zone boundary or fault line are reflected off the satellite back to ground receivers. Any ground movement is instantly detected and people warned.

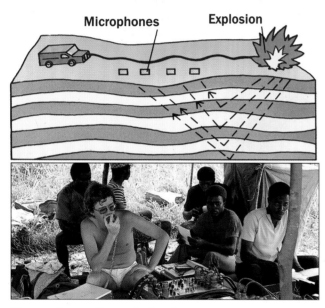

Measuring reflected shock waves

Another way of finding out what is beneath the earth's surface is to take a core sample. A hole is drilled into the earth. Then a special hollow bit (the part of the drill that cuts) is fitted to the drill. This bit cuts a cylinder of rock called a core sample which can then be brought to the surface and studied.

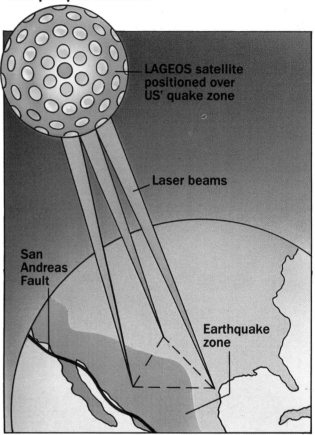

Laser-beam satellite

Drilling for core samples to find minerals

GLOSSARY

Anticline
The type of fold in which layers of rock are domed upward.

Core
The central part of the earth. The inner core is solid metal; the outer core consists of molten iron.

Crust
The outermost layer of the earth, consisting of the rocks that form the land and the ocean floors.

Fault
A crack in the rocks that form the earth's surface. Blocks of rock separated by faults may move vertically or sideways.

Fold
A bend in layers of sedimentary rock, caused by earth movements.

Guyot
A flat-topped underwater mountain, formed when a volcano "sinks."

Igneous rock
The type of rock that forms when magma becomes solid.

Lava
Molten rock that pours from a volcano.

Magma
Molten rock that lies just below the earth's crust.

Mantle
The layer of the earth between the crust and the outer core.

Metamorphic rock
The type of rock that forms when temperature, pressure or other forces change other rock types in form or in chemical composition.

Mid-ocean ridge
A long ridge, resembling an underwater mountain range, along the center of an ocean bed. It is where magma comes up from beneath the crust, creating new ocean floor and pushing the halves of the old floor apart.

Plate
One of the large slabs of rock that form the earth's crust. The plates move very slowly as they float on the molten magma beneath them. Continental plates are five or six times as thick as oceanic ones.

Plate tectonics
The theory that crustal plates move causing continental drift, earthquakes and volcanoes, and result in the formation of mountains.

Rift valley
A wide valley that is formed when layers of rock move downward between two faults.

Sedimentary rock
The type of rock that forms when sediments (of particles such as soil and sand) form layers under water and are squeezed by pressure to form rock. Some sedimentary rocks are formed from deposits of chemicals.

Syncline
The type of fold in which layers of rock dip downward.

Trench
A deep gorge along the ocean floor where one crustal plate is submerging under the edge of another one.

Vent
A hole or split in the earth's crust through which volcanic gases, ash, or molten lava escape.

INDEX

A
acidic rocks 14
anticlines 16, 31
atmosphere 6, 7, 15
atolls 21

B
barchans 22
basic rocks 14
block mountains 18, 19

C
caves 26, 28
coastlines 26, 27
cones 12, 13
continental drift 8
convection currents 8
coral islands 21
core 6, 7, 31
core samples 30
crust 5-9, 12, 14, 16, 20, 31

D
deserts 22, 23
dome 16
dome mountains 18, 19
dreikanters 22

E
earthquakes 5, 7, 10, 11, 16, 17, 30
epicenter 10
erosion 5, 15, 16, 18, 19, 22, 24-28

F
faults 16-19, 31
focus 10, 11
fold mountains 18
folds 16, 18, 19, 31

G
glaciers 24, 25
granite 14
guyots 21, 31

I
ice 5, 22, 24, 25
igneous rock 14, 16, 31
inselbergs 22

L
lava 12-14, 31
limestone 15, 28

M
magma 5-7, 12-14, 18-20, 31
magma chamber 12, 19
mantle 7, 9, 12, 14, 31
Mariana Trench 20
Mercalli scale 11
metamorphic rock 14, 15, 31
meteorites 28, 29
midocean ridges 20, 31
Mount Kilimanjaro 19
Mount St. Helens 13
mountains 5, 9, 12, 16, 18-19

O
oceans 8, 20, 21

P
Pangaea 8
pedestals 22
plates 5, 6, 8-10, 12, 13, 16, 18, 20, 21, 31

R
rain 5, 6, 15, 28
reefs 21
reverse faults 17
Richter scale 11
rift valleys 31
rock 5, 7, 14-16
rock cycle 14

S
salt-plugs 16
San Andreas Fault 16, 30
seafloor spreading 20
sedimentary rocks 14-16, 31
seif dunes 22
seismographs 11
shock waves 10, 11, 30
silica 14
strata 16, 19
synclines 16, 31

T
transcurrent faults 17
tremors 11
trenches 20, 31
tundra 25

U
underground erosion 28
undersea volcanoes 13, 20

V
valleys 24, 25
Vesuvius 13
volcanoes 5-7, 9, 12-14, 18, 20, 21

W
water 22, 24, 26
waves 26, 27
wind 5, 15, 22

Photographic Credits:
Cover and pages 16-17: Robert Harding Picture Library; pages 5, 19 top and bottom, 22-23, 25 top, 26-27 all and 28-29 all: Spectrum Colour Library; pages 9, 11 and 13: Frank Spooner Pictures; pages 10 and 21: The Hutchison Library; pages 14, 25 bottom and 30 top and bottom: J. Allan Cash Photo Library; pages 15, 18 and 24: Eye Ubiquitous; page 23: Science Photo Library.